Dedicated to my bee mentor Stuart Ratcliff
@Ratcliff_horizontal_hives

At last, Bee Curious?

ISBN 978-1-914934-69-8

Published by Northern Bee Books, 2023
Scout Bottom Farm
Mytholmroyd
Hebden Bridge HX7 5JS (UK)

Illustrated by Sarah Dolislager

WHY BEE CURIOUS?

Your interest in bees may be inspired by the humble honey bee and how they live together as individuals and yet act as one organism. Or you may want to help native bees thrive and multiply. Or you may want to harvest honey and beeswax for your own consumption or as part of a side hustle. Whatever the reason or reasons, this workbook is designed to help give you some basic information to help you decide how you might engage with native bees or keep honey bees.

This primer is intended to introduce you to the most essential information you need to help you decide if you want to pursue the path of gardening for native bees and/or adopting your own bee hive.

SPOTLIGHT: A BRIEF HISTORY OF BEEKEEPING

A Brief History of Beekeeping
Stuff You Missed in History Class
(May 11, 2020)

For more information on the history of Beekeeping, this succinct podcast tells more.

We are going to explore many different approaches to beekeeping. You can provide pollinator plants and enjoy observing a variety of insects and birds visiting your garden. Or you may host native bees in "houses" or by providing other bee-friendly spaces in your garden and perhaps a water source. Beekeeping is more intentional, by taking on the task of managing a bee colony.

Today beekeeping ranges from natural beekeeping with little interference to managing bees like livestock and moving them to various pollen sources. There is a middle way that involves taking steps to prevent dangerous Varroa mite and other pest infestations, regular hive inspections, and harvesting honey in modest amounts.

You will need to find what style suits you best. I've often heard the joke, that for every 10 beekeepers you'll hear a dozen opinions. After some experience you will have your own opinions. This is one of the aspects of beekeeping that keeps it interesting.

POLLINATORS

Pollinators are in crisis and that means we are too. Bird and insect populations are crashing in most parts of the developed world. The pressures are from loss of habitat, increased chemicals in the environment, predation from domestic pets, and other factors.

Disappearing bees, other insects, and birds makes life more difficult for everyone. About one in three swallows of food or drink are dependent on pollinators. While many grains can be wind pollinated, the fun food we love needs a "third party" to pick up pollen from one plant and carry and brush it off on another plant.[1]

1 *Attracting Native Pollinators*, the Xerces Society. Storey Publishing, North Adams, MA, 2011.

Scientists have measured the difference between the productivity of orchards and field crops with honey bees brought in, versus those without honey bees, and the difference can be up to one-third higher yield. Bird and insect pollinators are part of the biodiversity of a healthy planet.

EVOLUTION OF BEES

Flowers and bees evolved together. One hundred and thirty million years ago in the dinosaur epoch, flowering plants made up 80 per cent of plants on earth. As flowers became more complex and diverse so did the pollinators that collected their nectar. The first bees distinguished themselves from predatory wasps by sprouting feathery hairs on their legs and abdomen that can carry pollen back to their nest to feed their offspring and incidentally pollinate flowers. Flowers evolved ways to attract bees through color, scent and petal shapes. Bees developed long tongues to dip into nectar and evolved special stomachs for carrying nectar. Honeybees developed hive structures for honey and storing pollen to feed brood.[1]

The record of human management of bee hives dates back to the Ancient Egyptians where they used honey to pay their taxes, and for medicinal purposes. Honey was found in the pyramids (it's true–properly stored honey never spoils) but it may be an urban legend that pots of honey were found and tasted by early archeologists.[2]

---

1 *The Little Book of Bees*, Hilary Kearney. Abrams, New York, 2019.

2 *Plan Bee: everything you ever wanted to know about the hardest-working creatures on the planet*, Susan Brackney. Penguin Group, New York, May 2009.

For most of history, humans have been raiding wild bee hives often to the detriment of the bees. People built hive spaces into walls in Greece, Turkey and Iran. People hunted for hives in the treetops throughout Europe's forests. Eventually, woven skep baskets also became popular, with the drawback that to harvest the honey you have to destroy the hive. For much of the human experience we have valued honey and sought it out, but it is not until the 1800s that people begin to manage bees in what we would call a modern way.

The Swiss naturalist François Huber's discovery of "bee space" revolutionized the relationship between bees and humans. He was a gentleman-scientist in the same generation as Charles Darwin. He shared his bee behavior discoveries through letters to fellow scientists. Ultimately, this information flowed to Lorenzo Langstroth in the United States and he developed what today is most commonly known in the U.S. as a "typical hive"–that is how popular and utilitarian the Langstroth hive has become.

Langstroth's hive also enabled the industrialization of bees with hives moved from North Dakota then California then Florida and back to cold storage in North Dakota in one calendar year.

The British Standard Commercial hive was developed by Samuel Simmons and is also called "The National Major" hive.

These hives are relatively inexpensive with easily replaceable parts. You can find another beekeeper with the same equipment if you need some in a pinch. There are other hive designs, including the horizontal hive using Langstroth frames, and the Top Bar hive with removeable, but free-form comb. Just over 170 years ago, the Langstroth hive revolutionized bee management. Now beekeepers could harvest wax and honey without destroying the hive.

Langstroth Hive

UK Standard Hive

SPOTLIGHT: HUBER'S LEAF HIVE, THE FIRST MOVEABLE HIVE THAT ALLOWED FOR SOME OBSERVATION.

Huber's accomplishments are all the more impressive because he made these hive observations while blind. "When we have all the combs before us, ...we see how abundant the provisions are and what share of them we may take away.' Huber wanted beekeepers to see into the hive, but was wary of intervening too heavily.

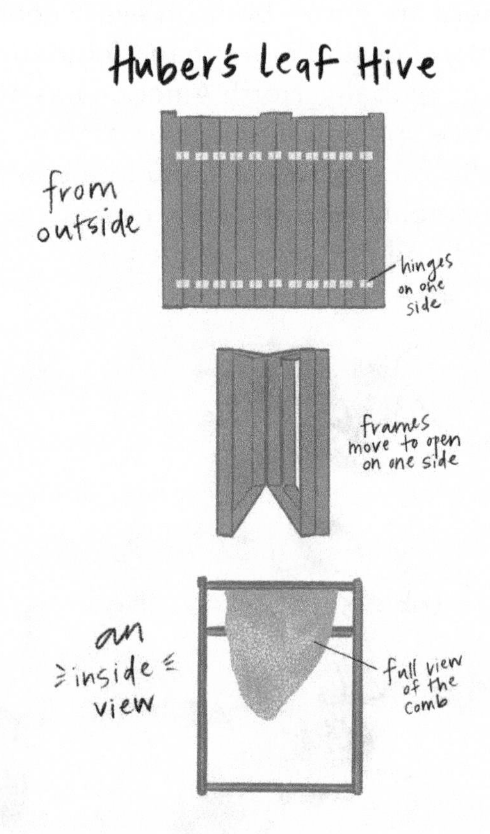

He urged moderation in the harvesting of honey and wax, and argued that to compensate for this moderation every means must be employed to promote the multiplication of bees. We should take a little; and do everything we can outside the hive to help them thrive."[1]

---

1 Helen Jukes, *The Honeybee Heart Has Five Openings*. New York: Pantheon Books, 2018, p. 60

NATIVE BEES IN YOUR GARDEN

If you spend time in your garden, you will begin to notice a wide variety of pollinators from flies to hummingbirds. There are native bees on every continent except Antarctica and more than 20,000 different species of bees globally. North America has an estimated 3,600 native species, and due to its varied ecosystems, in California there are 1,600 bee species.[1] The most commonly recognized varieties are bumble bees. Others might be mistaken for flies.

Natives are mostly solitary bees; each female constructs and provisions her own nest without any help from others of her species. These solitary bees live for about a year,

---

1 *California Bees & Blooms*, Gordon W. Frankie, Robbin W. Thorp, Rollin E. Coville, and Barbara Ertter. Heydey in collaboration with California Native Plant Society, Berkeley, 2014.

although we only observe them at the adult stage, which lasts about three to six weeks. These insects spend their early months hidden in the nest, growing through their egg, eating the pollen provided by mother at the larval stage and then pupating and emerging as a fully formed adult.

You can encourage native bees to thrive by providing them suitable habitat to breed and hibernate. For many native bees they need only undisturbed, bare ground. Ground nesting bees can be difficult to observe because the activity is mostly below ground. Bumble bees build their nests in old animal burrows and other holes. You can create artificial nests for bees. Nest blocks are holes drilled in a block of wood, or stem bundles with hollow reeds or bamboo. There are some excellent examples in the Xerces Society Guide, *Attracting Native Pollinators*.[1]

Popular plants with bees include sunflowers, daisies, dandelion, borage and verbena. Thistles such as artichokes, mint, peas or beans provide flowers for bees and produce for the gardener. Asters provide blooms from late summer into fall.

---

1 *Attracting Native Pollinators*, Eric Mader, Matthew Shepherd, Mace Vaughan, Scott Hoffman Black, Gretchen LeBuhn. Storey Publishing with the Xerces Society, 2011, pp 138-146.

SPOTLIGHT: DO HONEY BEES COMPETE WITH NATIVE BEES?

The late Extension apiculturalist Dr. Eric Mussen at UC Davis said no. "Honey bees should be solicited for restoration areas, not banned."[1] Indeed, they may be essential in re-establishing native plants that will support the return of native bees.

HUMMINGBIRDS OF NORTH AMERICA

Hummingbirds, with their long slender bills and a visual spectrum that includes red (unlike most bees), are able to feed from tubular flowers. They can feed from other flowers, but these types have fewer competitors.

Hummingbird feeding on an Amistad Salvia

1 "Honey Bees: Should They Be Banned From Native Plant Restoration Areas?" Kathy Keatley Garvey, *Bee Culture*, August 2022, p 69.

In Europe, pollination is by wind or insect, but in North America there are more than a dozen species of hummingbirds that pollinate plants.

Hummingbirds are mostly solitary coming together to mate. They don't fly together in migration. If you have a hummingbird feeder you've probably observed the competition for access to the feeder. Their aerial antics are dazzling.

Sage and salvia plants offer beautiful and hardy plants for the garden that become favorites of hummingbirds. Bushes such as toyon and elderberry provide flowers and then berries for other birds. Some birds eat bees but do not worry–the birds will not eat enough to impact the health of hive.

BUTTERFLIES AND MOTHS

Butterflies have long tongues that enable them to feed from tubular flowers, but they cannot hover like a hummingbird. They need a platform to land on with room for their wings. Butterflies also need the preferred plants for their larvae to feed on, as well as places to cocoon and then as adults they seek nectar from a variety of flowers. The Anise Swallowtail prefers plants in the carrot family such as parsley and bulb fennel. Western

Tiger Swallowtails like sycamore trees. The Painted Lady likes the mallow and aster families.

Another option is to cultivate a moon garden to attract moths. There are far more varieties of moths than butterflies but we are less likely to observe them as they are mostly nocturnal. Moon gardens offer blooms that are visible at night and attract moths. Plants include narcissus, angels trumpets, and creeping phlox.

SPOTLIGHT: ENDANGERED MONARCH BUTTERFLIES

The iconic Monarch Butterflies can feed from the nectar of many plants but rely on milkweed to lay eggs and feed their larvae. Many non-profits are working to recover Monarch Butterflies mainly by restoring drifts of milkweed. You can participate through one of the many programs on offer, such as Saveourmonarchs.org.

PLANTING FOR POLLINATORS

The best first step to ready your garden for pollinators is to STOP using synthetic chemicals in your garden. Pollinators are very sensitive to chemicals and the best chance of success is to focus on soil health and organic gardening. There are natural fertilizers such as fish emulsion or comfrey leaves soaked in water. While you are pursuing soil health you may see some pests–such as aphids–flourish. Resist quick fixes through pesticides.

Ultimately your garden will rebalance with the good bugs outweighing the bad. Similarly, when you commit to using natural products such as bark for pathways and mulch, more mushrooms pop up. This is also a sign of health.

Diversity is best for pollinators and makes a more beautiful garden. Attract more pollinators to your neighborhood by reducing lawn areas or swapping out the "boring" plants that don't support pollen-rich blooms.

SPOTLIGHT: CREATE A BEE OR BUTTERFLY BUFFET

You don't need a big garden to start. A few plants in a pot on your patio or balcony can attract butterflies and bees. Plants with many petals, such as peonies, have traded pollen for petals. Enjoy the ornamentals but save room for the pollinator favorites, such as foxglove, viola, or campanula. Please use a peat-free compost.

Also, consider the timing of the blooms to extend your garden's attractiveness from March to November. Planting in drifts of species makes for easier management and greater attraction. Look hard at the potential for more plants along alleys or in overlooked parts of your garden or shared spaces.

Native plants are best because they tend to be the plants local native bees depend upon and require less water. Relying on the nursery tags that boast "bee friendly or butterfly friendly" is not as reliable as doing a little research about your garden conditions (water availability and sunlight levels) and a good resource book. Gardening with natives also is a permission slip to let your garden look a little more relaxed. Prune lightly.

Your garden can also gradually evolve into a pollinator haven. Every spring and fall you can plant more plants as your time and budget allow.

SPOTLIGHT: THE FLOWER AS NATURE'S PUZZLE BOX FOR BEES

"...in visiting 1,000 flowers, the bee has to work 1,000 floral "puzzle boxes" whose mechanics can be as complicated as operating a lock, and no two flower species are quite alike in the mechanics that have to be learned to gain access to their contents. While flying through a flower meadow, the bee is constantly bombarded with stimuli (color patterns, scent mixtures, electric fields) from multiple flowers of several species per second, requiring the bee to pay attention only to the most relevant stimuli and to suppress the rest. Between visits to 1,000 flowers, the bee may have to reject 5,000 other flowers that either are unfamiliar or have been found to be poorly rewarding, or only rewarding at a different time of day."
The Mind of a Bee, Lars Chittka. p. 8

THE EUROPEAN HONEY BEE

Bees are one of those rare creations in nature known as super-organisms, that is they are a group of organisms that behave in some respects as a single organism. While a hive is a complex system commonly consisting of more than 50,000 individuals, it can also behave as if it were one organism or ecosystem. The ecosystem that the colony is a part of is as important as the physical hive. The Germans call the hive and the ecosystem it serves "bien". When the bien is thriving, all is well.[1]

The members of the colony include the queen who is the mother of all of the bees in the hive in a nuc and potentially unrelated in a package of bees. A "nuc" is so called because you are purchasing the nucleus of a hive, that is the queen, a couple of frames of brood and a few thousand worker bees. Whereas, in a package of bees the queen comes in a separate cage so she has time to communicate with the hive through her pheromones. She sets the quality of life in the hive. The workers will continue to support her while she is laying well. If she begins to falter, the workers will feed a few female bees royal jelly and raise a new queen. This process is called supersedure. A strong queen will live several years. During an inspection keep an eye out for the queen. Many a queen has been crushed by a rushed beekeeper clumsily replacing the frame she is on. Also,

---

1 *What Bees Want*, Susan Knilans and Jacqueline Freeman. New York: The Countryman Press, 2022.

keep the frame over the hive so if she does fall off, she falls into the hive.

The vast majority of the bees are workers. They are female and can fly and sting when mature. The drones on the other hand are male, cannot sting, and are slightly larger with eyes that take up most of their head. They are often mocked for their uselessness around the hive; however, the drones go out each day to fly in the neighborhood spot in the hopes of finding a queen to mate with and play an important role in maintaining bee diversity.

There are different types of European honey bees: Italian, Russian and Carniolans are the most common. Different beekeepers have preferences and you can spend some time reading about the types before you decide. There are also people breeding hybrid bees with the aim of creating queens and colonies that will behave more aggressively toward pests or may show resistance to Varroa.

SPOTLIGHT: WHAT IT'S LIKE TO BE A BEE

"I invite you to picture what it's like to be a bee. To start, imagine you have an exoskeleton–like a knight's armor. However, there isn't any skin underneath, your muscles are directly attached to the armor. You're all hard shell, soft core. You also have an inbuilt chemical weapon, designed as an injection needle that can kill any animal your size and be extremely painful to animals a thousand times your size–but using it may be the last thing you do, since it can kill you, too. Now imagine what the world looks like from inside the cockpit of the bee.

"You have 300-degree vision, and your eyes process information faster than any human's. All your nutrition comes from flowers, each of which provides only a tiny meal, so you often have to travel many miles to and between flowers—and you're up against thousands of competitors to harvest the goodies. The range of colors you can see is broader than a human's and includes the ultraviolet light, as well as sensitivity for the direction in which light waves oscillate. You have sensory superpowers, such as a magnetic compass. You have protrusions on your head, as long as an arm, which can taste, smell, hear, and sense electric fields. And you can fly. Given all of this, what's in your mind?"

The Mind of a Bee, Lars Chittka. Princeton: Princeton University Press, 2022. p.3

For more information, listen to:

The Mind of the Bee - Lars Chittka; 2 Million Blossoms - The Podcast (June 29, 2022)

BEE BEHAVIOR

SEVEN JOBS OF WORKER HONEY BEES

The all-female workers' hive assignments change throughout their life. The first 15 days are in the hive as their wings develop. The very first task after emerging is to clean up their own cell to make it ready for a new egg.

They also nurse the brood (larvae, pupae and young) and some will be charged with housekeeping and undertaker. Bees are fastidious and keep the hive clean from bee parts and debris. As the bee matures, its wax glands develop and it's able to secrete wax and build comb, cap pupae and ripen honey cells. These architectural chores also include filling cracks in hives with propolis. A select dozen will groom and feed the queen.

When they age further, they may be tasked with cleaning their fellow workers when they return from foraging for pollen or nectar, storing it in cells for later. Some of these bees will mix raw nectar with digestive enzymes and then fan to reduce the moisture and create honey. The final stage when the worker has a working stinger with venom, they will take on the most dangerous tasks of forager or guard. The forager is tasked with scouring a 3-mile radius for nectar and pollen. Once its honey stomach is full and its pollen baskets are full they will return to unload and then go out again until the sun goes down. The guard bees are fewer in number but charged with keeping intruders out of the hive. Bumble bees, wasps, and honeybees from other hives all must be barred from entering. If you are stung while you are working in the hive it is likely a guard warning you to go away.

WASHBOARDING AND BEARDING ON THE OUTSIDE OF THE HIVE

You may observe the workers on the outside of the hive on the front board moving back and forth seemingly without purpose. I call it line dancing but the official term is washboarding. There are several possible explanations but my favorite: it is the bees' way of expressing their joy of living.

On hot days you may see them bearding outside the hive, sometimes hanging down in a clump like a beard. This is a way of moving some of the hive members outside during hot temperatures to manage the inside temperature. Bees manage the temperature to maintain the hive between 95 and 96 degrees Fahrenheit (35 degrees Celcius). On hot days they will fan on the landing board to draw air through the hive (be sure to keep the vent open), and on cold days they will fan inside the hive to keep the colony warm.

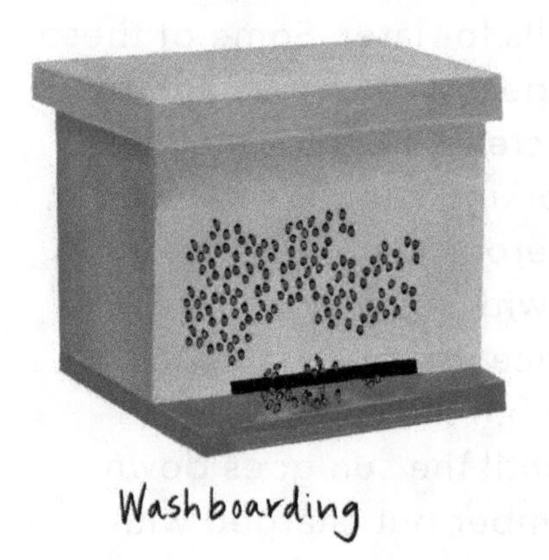

Washboarding

Bearding

THE WAGGLE DANCE

Worker bees have many ways of communicating with the other workers in the colony including a complex pheromonal communication system used to relay messages. For example, one scent used to indicate danger smells like bananas, which is why some people advise not eating bananas on the day you are working in your hive. The most famous bee method of communication is the waggle dance. A forager uses this elaborate dance to indicate where she has found a good food supply or a new home.

FESTOONING

When you inspect your hive and slowly remove a frame, you will likely see bees festooning. Bees have six legs each and their legs are equipped with a hook. They are able to create a chain of bees, but to what purpose? No one is entirely sure. Is it a way to measure distance? Is it to control temperature or more efficiently create wax? Observe for awhile and draw your own conclusions.

STINGS

Honey bees are the only bees that will die when they sting you. When she stings you and the barbs catch in your flesh, part of her abdomen will be torn off and this will kill her. There is venom delivered with the sting and this is what causes varying degrees of reaction. Removing the stinger with your hive tool as soon as possible and applying ice will help lessen the potential swelling.

SWARMS

People find bee swarms to be frightening, but it is actually when the colony is the most docile. A swarm occurs when about 10,000 members of the hive plus the old queen fly in search of a new home.

While it might look chaotic, there actually is a method to the seeming madness. Often people will call a swarm hot line when the bees rest in a large cluster. If the beekeeper arrives in time, he or she can collect the swarm and introduce them to a new hive. If the bees identify a suitable home more quickly, then they will fly off to set up their colony in a tree hollow, a space in someone's attic or an empty hive that might have bait inside.

These are just some of the interesting behaviors you might observe if you become a beekeeper or spend time in an apiary. Many beekeepers find talking to the bees helps to build a relationship. Bees can recognize human faces, detect bombs, detect tuberculosis, and more. Their intelligence is generally under rated.

BACKYARD BEEKEEPING

Keeping one or two hives in your garden can be very satisfying by investing a little more time doing inspections, treating for pests, and making sure the colony has the frames they need to thrive. It may mean some supplemental feeding if the conditions require it. In exchange you can harvest several gallons of honey and some wax each spring and summer once the hive is established. Plus, you have the pleasure of learning more about cooperation in nature.

The hive lives much of its life without much concern for us. It is only because we have flooded our environment with chemicals and paved over so many of the places where bees once foraged, that they need our assistance to thrive. Honey bees kept primarily for the environment need very little assistance. Some people do not treat for Varroa mites or stop swarms, and they choose hive styles that require little maintenance; however, this path can be a frustrating experience with colonies absconding every fall and starting over each spring.

Thomas D. Seeley quotes Wendell Berry "We have never known what we were doing because we have never known what we were undoing; We cannot know what we are doing until we know what nature would be doing if we were doing nothing." (Wendell Berry, "Preserving Wilderness," 1987). He advocates for a more hands-off approach that is often called "Natural Beekeeping."

A YEAR IN THE LIFE OF A BEEKEEPER

In early spring when you pick up your bee package or nuc, your bee year will begin. You will want to provide sugar water (1 cup of sugar dissolved in 1 cup of water) for the first several weeks. The worker bees will need it to draw out the comb and whilst they get oriented to the neighborhood. The queen will begin laying eggs right away, the workers will continue their tasks. When the hive is thriving and can support drones, the queen will lay some. These cells will be larger so look for them as a sign of the colony on track when you are inspecting. Your bees were likely treated for mites before you collected them. In May and June, you can check for mites using the sugar roll or other test method and treat if needed. Watch the frames fill to add a super or two as needed. June is prime time for swarms if the bees feel crowded and begin looking for a larger abode.

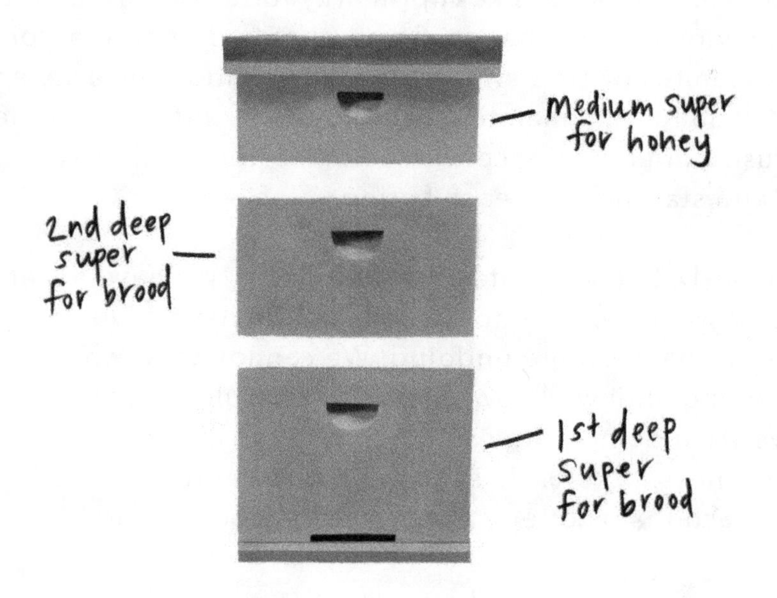

In summer, the "honey flow" will be going strong. Honey flow is when the nectar is most available and the bees produce the most honey. It starts around mid-May and continues till the first part of August. In July, begin treating for Varroa so the hive heads into fall and winter as mite free as possible. Observe your hive regularly and inspect every couple of weeks. Every time you inspect it takes the bees 3-4 days to repair your damage, so create a plan for inspections and keep notes in a notebook to make the most of each inspection.

In August and September, you can evaluate if the hive has enough honey for winter. If you have a Langstroth hive, consider leaving five full frames up to one full honey super. Better to wait a year to harvest honey than deprive the colony what they need to survive.

Don't remove a frame of honey until all the cells are capped with wax. Consider treating for Varroa once more in autumn. In suburban and urban neighborhoods bees will continue to find pollen and nectar.

Winter is a time not of hibernation, but of a more dormant state. Days are short and when it is cold the bees will stay in the hive helping to keep it warm and take short flights to poop outside the hive. Leave the hive alone and do not disturb. If they are not going to make it, there is nothing you can do, so improve their survival odds by leaving them alone. You can use this time to clean your equipment and think about the coming year. Consider expanding your garden.

EQUIPMENT TO GET STARTED

The basic gear you need depends on your comfort level with the bees and concern about getting stung. There are people who work with their bees without any body protection or smoker, but they are rare and move slowly and calmly and are not allergic to stings. Most new beekeepers are more likely to regularly inspect and do other bee tasks if they use a hat with veil, a smoker, and possibly some gloves. It is normal to be a little nervous and clumsy as a beginner working with bees, and it is helpful to have the gear handy if you need it. It is not uncommon to get stung even with protection, so if you are highly allergic you need an epi pen handy or consider another hobby. You'll notice as you watch videos that people range from tank top and shorts to fully suited. The other essential start-up purchase is a hive tool, a kind of Swiss Army knife for the beekeeper.

Some say the smoker masks the danger signals from the bees or the queen and so calms the bees. Others say the smoke tells the bees there is a fire coming and they focus on gorging on honey and leave you alone.

Either way, use non-toxic fuel for your smoker–such as old blue jeans–and do not depend solely on it. Keep listening to the hive. You'll notice the buzzing gets louder when they feel threatened. The guard bees will bump you. You can use smoke or you can step a few yards away and wait for them to calm down. Take that time to calm yourself with a deep breath and recentering on the hive.

SELECTING YOUR HIVE TYPE

Recall that a hive's needs are simple: a dry chamber protected from the elements with enough space to rear brood and store honey. In the wild they select space with ventilation and a defensible entrance. A swarm collector will tell you that bees sometimes select improbable places to build a hive, so don't stress. Your selection of equipment for your hive should align with your goals and ability.

The Langstroth hive is typical in the U.S. and the National Major hive is the UK standard hive. Both are managed in boxes typically having 8 or 10 frames per box. The brood box is commonly the bottom box and the surplus honey is collected in the boxes above (called "supers"). A queen excluder is often used to keep the queen in the bottom box. The hive components are widely available and interchangeable. Beekeeping information is mostly oriented to this type of hive and more mentors will be available. It is also widely seen as the best for maximizing honey production and Varroa treatments are primarily designed for a Langstroth hive.

The Horizontal or Long Langstroth hive makes beekeeping available to people with impaired mobility or limited lifting ability. The hive uses standard Langstroth frames, inside a rectangular wooden box raised on legs. The beekeeper can manage it and the biggest "lift" is one frame at a time. It is the size of a smaller Langstroth so the beekeeper does need to manage the frames to ensure the bees have the space they need.

The top bar hive is also more accessible to people with impaired mobility or limited lifting ability. The hive consists of a long box (3 to 5 feet) with straight or slanted sides. Bars are placed across the top of the hive and the bees build a single comb down from each bar. The bars fit tightly across the top to create a solid roof and the width of the bars creates the "bee space" between the combs.

Some kind of roof protects the hive from the elements. The bees typically keep the brood on one side and the honey on the other side. Honey is harvested by crushing the comb and straining it. It destroys the comb and can't be returned to the hive after harvest.

SPOTLIGHT: ALTERNATIVE BEEKEEPING

The interest in natural beekeeping pairs well with horizontal and top bar hives. *Keeping Bees With a Smile* by Fedor Lazutin with Leo Sharashkin is a guide for independent-minded beekeepers who want to keep bees without treating them with chemicals, disrupting their homes and unnecessarily intruding on their lives. *Common Sense Natural Beekeeping* is chock full of Kim Flottum's advice (with Stephanie Bruneau) and helpful illustrations. If you are intrigued with top bar hives then *The Thinking Beekeeper* by Christy Hemenway will assist you with solid, practical information to get started.

Horizontal Hive

There are many other types of hive containers and you may want to spend more time exploring the options before you choose. It is okay to get started with one type and then try other options over time.

The hive is unlikely to produce enough honey in the first year to share with you, so you don't have to worry about honey extractors right away. Just know this can be done in a low-tech way by hanging a frame over a pan to a honey extractor that spins out 2 or 4 frames at a time. Also explore sharing equipment with others.

THE BEST SPOT FOR A HIVE

The best spot for a hive is nearby where you will see it daily. When you are a beginner beekeeper you need to spend time observing the hive. Are they regularly returning with pollen legs, are there new foragers orienting in the late afternoon? If you have to climb stairs or a ladder to reach the hive, you are less likely to observe or conduct inspections. It is helpful if you can keep your tools close by also.

Ideally you can maintain the vegetation around the hive without much trouble. Provide a source of water, although they will also seek it out and may select another source over the one you provide. My bees prefer my neighbor's fountain. Maintain water year-round in case your neighbor turns their fountain off. Bees may also seek out different water supplies to source a variety of nutrients.

Hive stands are helpful for keeping skunks and mice out of the hive. You can buy them or make them with cinder blocks and pallets. If you decide to start with two hives, consider placing them 50-100 feet from one another. This will make it less likely for them to intermingle. If they need to be closer then have the entrances face different directions. Beekeepers find success with entrances that face East or West.

Think about year-round hive access, placing it near a creek might be attractive in the summer but not a good choice if there is any chance of flooding. Wild fires are also devastating. Hives are difficult to move without losing some of the hive population. Pick a spot that has morning sun and afternoon shade and is out of the wind. Bees manage in many situations, so do your best to pick the best spot.

Talk to your neighbors. You can reassure them that the bees will keep to themselves. The area about 10 feet in front of the hive will be a flight path, so if you need to put a screen between you and your neighbors, the bees will adapt.

Let people know that there is not likely to be any honey the first year, but then plan to give some to your neighbors each year to ensure their enthusiasm. You'll be an ambassador for pollinators and you can help your neighbors understand the role they can play in recovering the ecological balance.

Two hives are permitted on lots less than 10,000 square feet. Four hives are permitted on lots from 10,000 to 20,000 square feet. This is a typical county regulation for apiaries: The minimum lot size for beekeeping is 5,000 square feet. Six hives are permitted on lots greater than 20,000 square feet. The county agricultural commissioner may require registering your hives.

Most regions have a bee club or association. They provide a lot of local knowledge, like the best ways to keep bears from your hive. In the U.S., the local cooperative extension is likely to offer classes and expertise helpful to a beginner. If you can find a bee mentor to answer your texts or show you how to inspect a hive, then you are truly blessed.

BEE PRODUCTS

The most obvious product is honey. The colonies in your apiary will create honey that will be flavored by the most common sources of nectar. In most suburban and urban settings this will be a lighter colored honey that tastes most like wildflower honey.

If the source of honey is more than 20 percent of one crop, then it can be labeled as such, for example orange blossom or blackberry.

If honey is your passion, visit a place like the HIVE in Woodland, California and taste various honey harvested and begin to learn to taste the difference.

The oldest honey product beside the honey itself is mead. Mead is humankind's first alcohol. It is fermented honey with water and can be flavored by the type of honey used and other methods. (The May 12, 2020 episode of *Kiwimana Buzz Beekeeping* podcast is an excellent introduction to micro-mead making.)

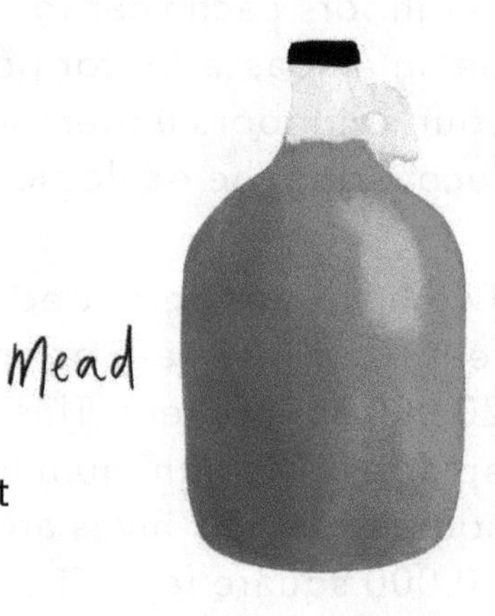

Beeswax is a wonderful product that can be used in a variety of products, most popularly candles. Other beekeepers harvest propolis and pollen for medicinal purposes. You can explore all of these opportunities, and remember Huber's admonition to be moderate in what you take from the hive, because the priority is maintaining a healthy colony.

CHALLENGES

Beekeeping does come with challenges and since 2006 the biggest challenge is Varroa mites. These are widespread among European honey Bees in North America and Europe. Weak colonies can quickly become overwhelmed because the mites lay eggs with the bee brood. Some swarms are probably the result of infestation as it can break the mites' reproduction cycle. There are chemical treatments available and there are variety of opinions about the most effective way to manage Varroa.

Another common pest is hive beetle. It can ruin a honey harvest and may prompt a colony to abscond (disappear in the autumn). A healthy hive will keep the beetle population down. Another pest that a healthy hive will manage are wax moths. It is helpful to manage your frames after harvest to avoid infestation by freezing older frames and storing in airtight containers.

The best defense for all of these pests is a strong healthy colony. Some people rely on supplemental feeding to keep a hive strong, but the best way to maintain bee health is plenty of nectar and pollen from flowers, shrubs and trees.

There are also pests that like the bees or the honey. Skunks like to eat bees at night and the best defense is elevating the hive. Spiders will often spin a web on the hive to be able to catch a bee. Regular external inspection helps to keep them at bay. Ants are attracted to the honey. You set the hives legs in water to break the access. DO NOT use any ant poison out of doors. Bears are literally the biggest challenge. Most neighborhoods don't have this worry, but if you are in the foothills then know that, like Winnie the Pooh, the bears are keen to get some honey and while the bees might sting them it is not enough to stop them breaking apart the hives. Electric fencing can provide hive protection.

MORE RESOURCES FOR THE BEE CURIOUS

Time and money are always somewhat limited, so I am recommending the best of resources in a variety of formats so you can find a path for your curiosity and need for bee knowledge. I am aiming for the backyard beekeeper. If you want to manage bees commercially, you will want to take a deeper dive. Here's where you can start.

BOOK

The Backyard Beekeeper, Kim Flottum. Massachusetts: Quarry Books. 2018. The author regularly updates this 230-page book with full-color illustrations. I have studied a lot of bee books and this is my favorite for the beginner. It has all of the relevant information and Kim Flottum is a reliable expert.

MAGAZINE

If you enjoy a more scientific approach, you may prefer *American Bee Journal*; however, for a large number of short articles on a great variety of topics–treating for Varroa mites to recipes– *Bee Culture* is the magazine for American beekeepers. In the U.K., *Bee Craft* is an award-winning monthly beekeeping magazine.

WEBSITE

There are a lot of websites with great information. It is hard to recommend just one. The award for worst design, best information goes to ScientificBeekeeping.com with Randy Oliver. For North American bird information, see allaboutbirds.org.

PODCAST

I've been learning about beekeeping from podcasts since I discovered the *Kiwimana Buzz Beekeeping* podcast from New Zealand. Alas, they stopped production after 160 episodes, but they are still available. *Beekeeping Today* is a podcast with regular new shows that frequently cover interesting topics. The hosts, Jeff Ott and Kim Flottum, keep bees in snowy climates, so some advice requires adaptation. Kim Flottum has a second podcast with Jim Tew, called *Honey Bee Obscura*.

YOUTUBE VIDEO CHANNEL

My go-to for explanatory videos is *Beekeeping Made Simple*. Laryssa in Hawaii answers a lot of my questions with great videography. It is also okay to type in your specific concern, such as "How to use Sugar Roll to check for Varroa mites in the hive" and watch several videos by different beekeepers.

INSTAGRAM

There are many accounts I like to follow so I get bees in my feed. @HoneyPieBees is a California backyard beekeeper who takes super photos. Posting to Instagram is a great way to raise awareness about pollinators and their importance.

PLANTS

The North American Native Plant Society will help to connect you to local resources because gardening is **local**! I am inspired every week by the BBC show, *Gardener's World*. Also visit local botanical gardens or demonstration gardens like the UC Davis's Bee Haven garden.

BEE CLUBS AND MENTORS

Your local beekeepers association is another place to find other beekeepers, potential mentors, and continue your education.

Finding a mentor is such a blessing. There is just so much to learn in the first couple of years and the confidence of a mentor can rub off on you. Yet most people don't find one and still persevere.

Printed in the USA
CPSIA information can be obtained
at www.ICGtesting.com
LVHW012336111223
766263LV00045B/1398

9 781914 934698